Let's Grow Green

EVERY ACTION COUNTS!

By Belinda Gallagher

Ruby Tuesday Books

Published in 2024 by Ruby Tuesday Books Ltd.

Copyright © 2024 Ruby Tuesday Books Ltd.

All rights reserved. No part of this publication may be reproduced in whole or in part, stored in any retrieval system, or transmitted in any form or by any means, electronic, mechanical, photocopying, recording, or otherwise, without written permission from the publisher.

Editors: Ruth Owen & Mark J. Sachner
Production: John Lingham

Photo credits:
Alamy: 24C (John Swithinbank); Dreamstime: 8 (Oleh Malshakov), 9TL (Photographyfirm), 14T (Yevheniia Ryzhova), 31BR (Kateryna Chyzhevska); Ruby Tuesday Books: 15, 17BL, 21B, 23T, 25BR; Shutterstock: Cover TL (Olga Danylenko/Dionisvera), Cover BL (Max_555/Soyka), Cover TR (Maples Images/Art_Pictures), Cover BR (Lois GoBe), 4T (Tatiana Gordievskaia), 4C (Peter Fleming), 4B (Maria Evseyeva), 5TL (Elena Masiutkina), 5TR (Mercedes Fittipaldi), 5CR (FeellFree), 5B (Kikujiarm), 6T (t.sableaux), 6C (Daisy Daisy), 6B (SviatlanaLaza), 7T (plazas I subiros), 7BL (miammaria), 7BR (Irina Starikova1811/Richard P Long), 8 (Tatevosian Yana), 9CL (neenawat khenyothaa), 9BL (LaNataly), 9TR (Jarno Holappa), 9BR (JulieK2), 10T (Kaca Skokanova), 10B (Graham Corney), 11TL (Sina Ettmer Photography), 11B (m.malinika), 12T (Alexandre Laprise), 12C (Wasu Watcharadachaphong/Suvorov_Alex), 12B (Eneems), 13T (Petr Bonek), 13B (Valerie Quemener), 14C (Saeedatun/New Africa/LuXpics), 14B (andres barrionuevo lopez), 16TL (Helen Kosareva), 16TR (Mehriban A), 16B (Mehriban A), 17T (Geshas/Fischeron), 17CL (bigacis), 17BR (irina2511), 18T (mehmetkrc), 18CL (Andrey Pristyazhnyuk), 18CR (Keith Hider), 18B (Gina Lee Rodgers), 19TL (poidl), 19TR (Shane Kennedy), 19BL (Jeremy Burnside), 19BR (Taviphoto/YK), 20T (Daniela Morgenstern), 20CL (Martin Bergsma), 20CR (FotoHelin), 20B (sjm1), 21T (Aleksandr Naumenko), 22T (aabeele), 22CL (Sergey Merkulov), 22CR (TunedIn by Westend61), 22BL (JossK), 22BR (Angeliiina13), 23C (Bogdan Wankowicz), 23BR (Denis Kuvaev), 24T (1000 Words), 24BL (Yuriy Balagula), 24BR (DavidEdwards8), 25T (Gail Johnson/telear), 26T (Pressmaster), 26CL (Yellow duck), 26CR (LariBat), 26B (Simol1407), 27TL (Ro_ksy), 27BL (Art_Pictures), 27R (mrivserg), 28TR (PaniYani), 28C (Joaquin Corbalan P/Nataliia Melnychuk/Steven Giles), 28B (Hipatia), 29C (Miriam Doerr Martin Frommherz), 29B (Robert Kneschke), 30TL (photka), 30BL (teatian), 30TR, 30BR (Cora Mueller), 31L (Photomaster), 31TR (Oleksandrum).

ISBN 978-1-78856-444-1

Printed in Poland by L&C Printing Group

www.rubytuesdaybooks.com

Note from the Publisher

Neither the publisher nor the author can accept legal responsibility or liability for any loss, harm or injury that may come about from following the instructions in this book. All activities should be carried out with adult guidance and supervision. Some activities involve being out of doors in public spaces. Children should be accompanied at all times. It is the parent's or carer's responsibility to ensure their child is safe.

CONTENTS

Let's Grow Green 4
Start with Some Recycling 6
Love the Soil 8
Start a Compost Heap 10
Save Some Water 12
Grow Your Food 14
Grow Plants from Leftovers 16
Protect Your Plants Naturally 18
Bulbs for Bees 20
Spread Some Pollen 22
Let Things Grow Wild 24
Plant a Tree 26
Harvest Your Seeds! 28
Glossary 30
Index 32

Staying Safe!

All the activities in this book are fun and easy to do. Be sure to ask an adult to help you with each one at every stage. Never go anywhere without your trusted adult. Wear old clothes and gloves when doing gardening activities, and make sure you are wearing clothes that suit the weather. Always wash your hands after touching plants and soil. Have fun!

Let's Grow Green

All living things need plants. Plants provide food and clean air. And they are homes for many kinds of wildlife.

Growing green is about how we grow plants and use our outdoor spaces – whatever their size.

Top Tip
You don't need a garden to start growing green. All it takes is a few small actions!

Bumblebee covered with pollen

Thistle flower

Did you know?
Bees, wasps, butterflies, beetles and other insects carry **pollen** from flower to flower. This helps flowers produce **seeds**. An insect that spreads pollen is called a pollinator.

Scatter wildflower seeds in your growing space or in containers filled with soil. Bees and other insects feed on **nectar** and dusty pollen from wildflowers.

Rainwater is natural and costs nothing. Catch rain in a water butt or buckets and use it to water outdoor plants in dry weather.

Water butt

Sometimes people use chemicals to kill wild plants known as **weeds**. These chemicals poison the soil and harm wildlife. Ask to keep your outdoor space free from weedkillers.

Try pulling up weeds and their roots by hand.

To start Growing Green, here are some tools you may need.

Watering can

Trowel, hand fork, and small spade

Gardening gloves

Care for a Plant

Ask an adult if you can buy a plant to care for. It can be an indoor or an outdoor plant.

Check the label on the plant to find out what conditions are best for it to grow in.

Some plants like lots of sunlight, while others prefer shade. Check how often the plant should be watered.

Cactus

Rosemary plant

Start with Some Recycling

Growing green means reusing and recycling as much as possible. As you recycle, think about what could be useful for growing green activities.

Seedling

Egg carton

Toilet roll tube

Fill yogurt cartons, toilet paper tubes, and egg cartons with soil and use them for planting seeds.

Did you know?
Cardboard is natural, and it rots. So it can be planted straight into the soil when **seedlings** are ready to plant out.

Make plant pots from plastic bottles.

Plants aren't fussy about what they grow in! Look for things around the house that can be reused as fun planters.

Top Tip
Be sure to ask an adult to make holes in the bottom of a planter so spare water can drain out.

Hang old CDs and DVDs on branches. Flashes of sunlight from the CDs will scare birds away. This will stop them eating your seedlings, vegetables or growing fruit.

You could even make a CD scarecrow!

Recycled Plant Labels

Make name labels for your plants with recycled items from your home. Here are some ideas to try!

Wooden spoons
Paint wooden spoons and add the name of your plant. Let the paint dry and then varnish the spoons to make them waterproof. Push the spoons into the soil next to your seeds or plants.

Wooden clothes pegs
You can paint the pegs to make them more colourful. Write on the plant names with a waterproof marker. Clip the pegs to your plant containers.

Paint pebbles
Paint pebbles with plant pictures and add your plants' names. Let the paint dry and then protect the pebbles with a coat of paint-on varnish. Place your pebble markers close to your plants or where you plant seeds.

Love the Soil

The soil that covers the ground is a **habitat** filled with living things. Plants need healthy soil to grow. How can we give soil a helping hand?

Roots

Did you know?
Plants need water and **nutrients** to grow strong. They take in these things from the soil with their roots.

Fungi · Tiny animals · Microbes

Did you know?
The living things in soil can be killed by chemicals, so never use weedkillers!

Soil is packed with tiny animals, **microbes** and fungi. Most are too small to see. These living things feed on rotting plants and other dead stuff, and turn them into new soil.

Let worms do their magic! Earthworms burrow in soil, eating dead stuff. Their burrowing breaks up the soil and lets in air and water.

Worm castings are worm poo. They add nutrients to the soil that help plants grow.

Worm castings

In autumn and winter, let fallen leaves rot into the soil. They will feed new plants in spring.

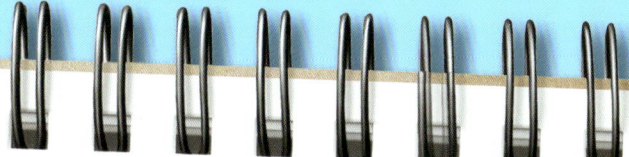

Clever Clover

If you have a patch of bare soil in your garden or school playground, sow clover seeds in springtime. When the clover flowers, it will provide bees and butterflies with pollen and nectar.

Clover flower

You will need:
- Clover seed from a garden centre
- A rake
- Water and a watering can

1. Remove any weeds by hand from the patch of soil.

2. Rake the soil to make it crumbly.

3. Gently sprinkle the clover seed over the raked soil.

4. Water the soil. Keep it moist for the next few days. The clover will sprout in as little as two to 10 days.

Clover leaves

Clover seeds

Once the clover flowers have died, you can dig the green clover leaves and stems into the soil. As they rot, they will add lots of nutrients to the soil.

Start a Compost Heap

One of the best ways to grow green is to make **compost** on a compost heap. You can do this by recycling food scraps and waste garden material.

Compost heap in a box

Waste goes in.

Compost is filled with nutrients. Plants love it, and it will help them grow.

The waste rots.

You can add your compost to flowerbeds or vegetable patches. You can also use it as soil in containers.

The waste becomes compost.

What is a compost heap?

It's a box or heap of natural, rotting waste. The waste slowly turns into brown, crumbly compost, or soil. Insects, worms, microbes and other tiny life forms will quickly move in to help break down the waste.

FROM THE KITCHEN

You can add vegetable peelings, apple cores, banana skins, eggshells, tea bags and coffee grounds.

FROM THE BACKYARD

Chopped garden waste

You can add things such as grass clippings, leaves, weeds, twigs and dead plants.

You can also add shredded paper and cardboard to a compost heap.

Let's Make Compost

It takes a little while for compost to form – so let's get started!

1. Put your compost box in your garden. You can also choose a corner of your garden and simply make a heap of waste.

2. Put food scraps in the kitchen waste bin. When the bin is full, empty it into your compost box or onto your heap.

3. Add garden waste to the compost box or heap.

You will need:
- A sheltered place in your garden or a compost box from a garden centre
- A small waste bin for the kitchen scraps
- Kitchen scraps
- Garden waste

4. Every other month, turn over the rotting material with a large garden fork. This adds air and helps things rot faster.

5. It can take a few months for the waste to turn to compost, but let nature do its magic! When the compost is ready, it should be dark and crumbly, with no remaining scraps or twigs.

Eat food → Food scraps → Make compost → Feed the soil → Grow plants → Grow more food

11

Save Some Water

Using water wisely is a big part of growing green. It takes energy and costs money to supply clean water to our homes, so try to use less tap water when growing plants.

Rainwater is best for plants, and it's free! Place buckets or bowls outside and see how much rainwater you can collect.

Wash fruits and vegetables in a bowl. Save the water and use it for watering plants.

Keep the water from cooking pasta and vegetables. Let it cool before watering plants.

Use a watering can – not a hose. Point the spout at the soil. Water will soak in around a plant's roots and not splash around and be wasted.

Top Tip
Grow plants, such as lavender and wildflowers, that can go without water for a long time.

Lavender smells wonderful, and it feeds bees.

Did you know?
Water is precious! Every drop we save in our gardens helps keep rivers and lakes filled with water for wildlife.

Make a Recycled Water Sprinkler

Don't throw away a plastic bottle. You can turn it into a recycled super-sprinkler to use for watering plants.

You will need:
- A plastic bottle with a top
- An adult helper with a small drill

1. Rinse the bottle (save the water) and remove the top.

2. Ask an adult to make several holes in the bottle top with a drill.

3. Fill the bottle with water you've recycled from the kitchen or garden.

4. Get busy watering seeds and plants!

Grow Your Food

It's easy to grow fruit and vegetables at home. If you have space, you can make a veggie patch. Or grow plants in containers. But why grow food when you can buy it from a supermarket?

Growing salad leaves in a veggie patch

Top Tip
Make recycled plant labels to mark which plants are growing in your veggie patch.

Supermarket food is often packaged in plastic. To keep our planet clean, we should use and throw away less plastic.

Packaged supermarket tomatoes

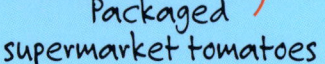
Homegrown tomatoes

Homegrown food is good for the environment. It doesn't need transporting from a farm to a shop. Sometimes farms use chemicals to kill weeds and plant-eating insects. Homegrown food is chemical-free.

Did you know?
When you buy packets of seeds, they have instructions on the back that tell you how to grow the plants. You can grow **herbs**, lettuce, tomatoes, peppers, carrots, radishes, onions, beans, peas, courgettes and many other foods from seeds.

Grow Your Own Strawberries

Start this activity in March or April to give your plants time to grow. You can buy strawberry seedlings online or from a garden centre.

You will need:
- A container such as a window box or bucket (with drainage holes in the bottom)
- 5 strawberry seedlings
- Peat-free compost, or homemade compost
- A watering can

1. Fill your container with compost.

2. With your hands, scoop out a small hole in the soil. Put a strawberry seedling into the hole, and scoop soil back around the plant so its roots are covered. Gently press down. Repeat for the other seedlings.

3. Put the container outside in a sunny spot, and water your plants.

4. Keep the soil moist by watering regularly.

5. After about four to six weeks, the plants will grow white flowers. Bees will visit the flowers to feed, and they will **pollinate** the flowers.

6. After a few weeks, a strawberry will grow in the centre of each flower.

7. When the fruits are plump and red, they are ready for picking and eating!

Grow Plants from Leftovers

Did you know that some fruits and vegetables can regrow from scraps? Don't throw all the scraps on the compost heap. Hold on to some and watch them magically turn into new food!

TASTY TOPS

Slice the tops off carrots and beetroots and put them in a bowl of shallow water. New leaves will soon appear. The leaves can be snipped off to add to salads.

After a few weeks, plant the tops in a pot of compost, and the leaves will keep growing.

SUPER STUMP LEAVES

Place the unwanted bottom stump of a lettuce or cabbage head in water. New leaves will grow!

GROW SPRING ONIONS

Put the bottom part of a spring onion (with roots) in a jar of water. New green stems, which you can eat, will grow from the stump.

Top Tip

Plant your spring onion stumps in a pot of compost and place them somewhere sunny. Snip off the green stems back to soil level when you want to eat them.

16

PINEAPPLE POT PLANT

When you eat a delicious pineapple, keep the spiky crown. Pull off any lower leaves and place the crown in water. Once roots grow, plant the crown in a pot of compost. You have a new plant to care for!

Keep your plant somewhere warm and bright.

Top Tip

Look at the food you eat and try experimenting with scraps. For example – sow seeds from lemons, oranges or limes in compost.

Slice-and-Grow Tomato

This fun activity grows tomato seedlings from a sliced-up tomato. It's that simple!

You will need:
- A ripe tomato
- A knife and chopping board
- A recycled plastic container
- Peat-free compost or homemade compost
- A small watering can, or a recycled water sprinkler

1. In spring, carefully chop a tomato into slices, as if you were going to eat it.

2. Place the slices of tomato flat on the compost in the container. Cover with a thin layer of compost, so you can't see the slices.

3. Water the compost to make it moist. Place the container in a warm, light place, and keep the compost moist.

4. After about two weeks, tiny seedlings should grow from the seeds in the slices.

5. Let the seedlings grow until they have two pairs of leaves. Then carefully plant each seedling in its own pot of soil.

Pass seedlings on to friends to care for. Who grows the biggest tomato crop?

Protect Your Plants Naturally

Some gardeners use poisonous chemicals to kill plant-eating slugs, snails and insects such as aphids and caterpillars. But there are lots of natural ways to protect your plants.

Aphids damage plants by sucking a sweet liquid called sap from them. Helpful ladybirds, hoverflies and lacewings eat aphids.

Aphids
Ladybird

Lacewing

Hover-fly

The helpful insects also feed on pollen and nectar. Grow flowers close to your vegetables and fruits to attract bug helpers.
(See page 23 for flower ideas.)

Nectar-rich marigolds

Leave an overgrown area in your garden where birds, toads and frogs can hide. They will hunt plant-munching slugs and snails.

Cut recycled plastic bottles down and place them over seedlings. This will stop hungry slugs and snails eating them.

Bring in the Birds

Feeding garden birds is fun and easy. And when you attract birds to your outdoor space, they will feed on plant-eating slugs, snails and insects. This is a great way to protect your plants and make your garden a natural, balanced habitat.

Put a feeding platform in your garden or hang feeders in places where cats can't pounce on the birds.

Bird food ideas:
- Plain cooked rice and pasta
- Chopped-up cooked potatoes
- Chunks of fruit such as apples and pears
- Seeds, suet balls, mealworms and bird peanuts

Bird feeder with peanuts

Caterpillar

Birds will come for your food...

...and help keep plant munchers under control!

Did you know?

It's good to offer food all year round. Nesting birds need extra food in the spring. Food is harder to find in the cold winter months.

Bulbs for Bees

Some plants grow from **bulbs**, which are an underground part of a plant. Flowering spring bulbs, such as crocuses, provide food for hungry queen bumblebees.

Bulbs stay underground all year round. They contain all the food that the plant needs to grow.

Snowdrops

Did you know?

Bumblebee queens wake up from their long winter rest in early spring, when there are not many flowers around. Flowers that grow from spring bulbs give bees nectar and pollen to eat.

Top Tip

Try planting snowdrops and crocuses in grass. Ask an adult to help you dig up a flap of grass. Push the bulbs into the soil, and then cover them again with the grass flap.

Flap of grass
Crocus bulbs

Crocuses

Even if you don't have a garden, you can plant bulbs in pots, window boxes and other containers.

Allium bulbs flower in early summer. Butterflies flock to allium flowers to drink nectar.

Alliums

Make a Bulb Lasagna

Do you like lasagna? This one isn't for eating! Instead, bulbs are planted in layers to give a pretty display that will flower from early spring onwards. Start this activity in the autumn.

You will need:
- A large deep pot or bucket with drainage holes
- Pebbles
- Peat-free or homemade compost
- A selection of bulbs (see below)

1. Add a layer of pebbles to the bottom of your container, then add a layer of compost.

2. Plant later-flowering bulbs such as alliums first. Place the bulbs on top of the layer of compost about 3 inches (7.5 cm) apart. Add a second layer of compost.

3. Plant a middle layer of bulbs – for example, tulips and daffodils. Add another layer of compost.

4. Finally, plant early-flowering bulbs, such as crocuses or mini daffodils. Cover with a final layer of compost. Keep the container in a sheltered spot.

5. Once the first shoots appear, keep the compost moist.

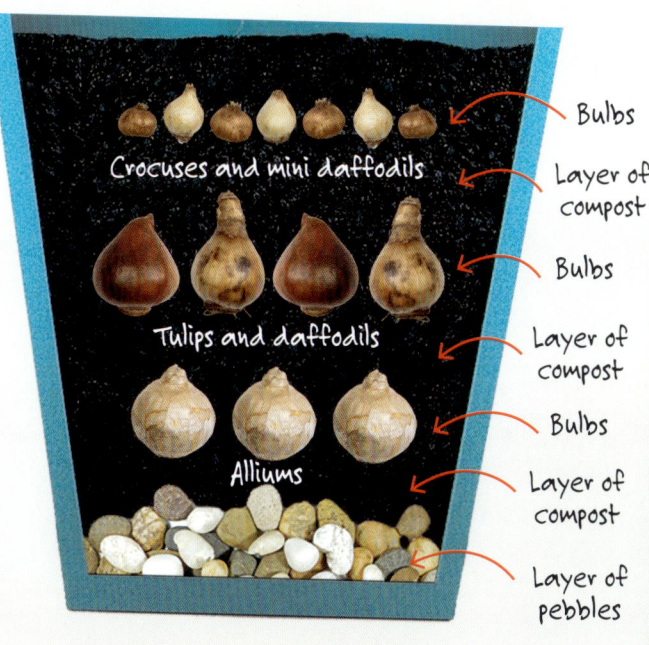

Did you know?

After a bulb's flowers have died, don't cut back the leaves for a few weeks. The bulb needs its leaves to make food that will be stored in the bulb for next year.

Spread Some Pollen

When we grow pollen-rich flowers, we give insects food. Then the insects help our plants produce even more plants to make our world greener and more beautiful.

Hummingbird hawk-moth

Did you know?
Butterflies and moths feed on sugary nectar with their long, straw-like tongues. As they feed, they help spread pollen.

Pollination in Action!
- An insect flies to a flower to feed.
- The insect's body becomes covered with pollen.
- The insect flies to the next flower of the same kind.
- Some pollen gets brushed off. That flower is now pollinated and can make seeds.

Pollen is carried from here . . .   . . . to here!

Create a mini herb garden in a container. Pollinators love the flowers of rosemary, marjoram and chives – and you can use the herbs for cooking!

Thick-legged flower beetle

Many kinds of beetles are also great pollinators.

Top Tip
Add a beetle habitat to your outdoor space with a simple bug hotel. You can make it from recycled cans stuffed with sticks, dried grass, and bamboo canes.

Bug hotel

Flowers for Pollinators

 Borage
 Cosmos
 Onion flower
 Oregano
 Lavender
 Nasturtiums
 Foxgloves
 Yarrow
 Hardy geraniums
 Poppies

Super Giant Sunflowers

Bees and hoverflies love sunflowers – and they are easy to grow from seeds. Start this activity in May to have gorgeous giant sunflowers in the summer.

You will need:
- A packet of giant sunflower seeds
- A trowel
- Peat-free or homemade compost
- A watering can
- Long sticks or bamboo garden canes

1. Choose a sunny patch of soil in your garden or school playground (with permission from a teacher). An area that's 1 metre by 1 metre will have space for 10 sunflowers.

2. Dig and turn over the soil to make it soft and crumbly. As you dig, add handfuls of compost. Remove any stones or weeds.

3. Next, make a hole with your finger. Drop in a seed. Plant more seeds, spacing them about 20 cm apart.

4. Cover the seeds with soil, water the soil well and wait!

Amazing things will happen in the soil.

Seed — Roots grow — Shoots grow

5. Use sticks or canes to support the flowers as they grow taller. You will soon have giant, buzzy sunflowers packed with nectar and pollen.

Sunflower seeds

Let Things Grow Wild

Step back and let patches of your garden grow green AND wild! Put the lawnmower away for a few weeks to let nature work its magic.

LET THE LAWN GROW LONG

You can still mow a strip through the lawn as a path. Or make patterns of short patches and longer, wild areas.

Top Tip

Let plants in your wild patches die back naturally. As they rot, the plants become food for worms and insects, and they add nutrients to the soil.

MAKE A MINI MEADOW

Plant a wildflower area and let weeds grow to create a mini meadow for pollinators.

DANDELIONS ARE SUPER WEEDS

Seeds

Dandelion flowers provide pollen and nectar for butterflies and bees. Their fluffy seeds are food for birds.

ALLOW MOSS TO GROW

Patches of **moss** stop the soil drying out. Birds gather moss to line their nests in spring.

COUNT THE BUTTERFLIES

Lots of butterflies will mean that the environment is healthy.

Make a Mini Moss Habitat

Moss loves the cool and damp. So make this mini habitat and place it in a shady part of your outside space. Your moss habitat will provide shelter, food and water for insects and other wildlife.

1. Half fill the container with soil. Push the plastic carton into the soil.

2. Cover the soil with moss.

You will need:
- A shallow container
- Peat-free potting soil or homemade compost
- Pincushion moss from a garden centre or garden
- A recycled, small plastic carton, such as a cream cheese container
- Pebbles, sticks, pieces of bark
- A watering can

What wildlife moves in? See if any frogs come by!

A moss habitat in an old frying pan

Pincushion moss

3. Fill the carton with water and add pebbles as mini stepping stones.

4. Water the moss and add sticks and other mini hiding places. The moss will grow, so if the weather is dry, water the moss to keep it damp.

Plant a Tree

Planting a tree is fun, easy and helps the environment in many ways. If you don't have a garden, a small tree can be grown in a pot. Or find out if there's a tree-planting project in your neighbourhood.

Terrific Trees

- When we walk or sit outside in nature, trees can help us feel calm.
- Trees release oxygen, the gas we need to breathe.
- Trees are home to birds, insects and other wildlife.
- The roots of trees hold soil together and stop it washing or blowing away.
- Fruit trees grow blossoms packed with pollen for bees, and give us fruits such as apples and pears.

You can grow a small potted tree on a doorstep, deck or balcony.

Lemon trees

Young apple tree

Top Tip

The best time to plant a tree is between November and March. Make sure the tree you choose will be the right size for your outdoor space as it grows.

Grow a Mighty Oak Tree

Acorns are the seeds of oak trees. Start acorn hunting in September and plant them in pots to grow your own mighty oaks.

You will need:
- 4 brown acorns with no parts nibbled from them
- A medium-sized plant pot with drainage holes
- 4 large plant pots with drainage holes
- Peat-free potting soil or homemade compost
- A trowel
- A watering can

1. Fill the medium-sized pot three-quarters full of soil.

2. Use your finger to make four holes, spaced apart. Put an acorn in each one.

3. Fill the holes and the pot with soil, and water well.

4. Place the pot outside. In spring, shoots will start to grow – these are oak tree seedlings!

5. Let the seedlings grow until the roots can be seen through the pot's drainage holes. Now it's time to carefully plant each one in its own pot of soil.

6. When the seedlings are about 12 inches (30 cm) tall, they can be planted in the ground.

Young oak tree

Be sure to give a young oak tree plenty of space to grow.

Ask others to plant trees, too!

Oak tree seedling

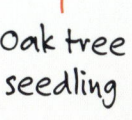

An oak tree can take 100 years to fully grow.

Harvest Your Seeds!

By now, you may have planted different fruits, vegetables and flowers on your growing green adventure. Once they have grown you can gather seeds for next year, and to share with neighbours and friends.

Dill seeds

As flowers die, you will see seeds forming in the dead flowers or in seedheads.

Poppy seedhead

Seeds

Dead flower

Sunflower seeds

Let the flower or seedhead dry out for a few weeks. Then collect the seeds.

Top Tip
Spoon the seeds from a homegrown tomato onto some kitchen towel. Let the seeds and kitchen towel dry out. Cut the kitchen towel into sections and save them.

Dried seeds

Cherry tomatoes September 2024

The pieces of kitchen towel with seeds on them can be directly planted in compost next spring.

Use recycled paper and some double-sided tape to make little envelopes for storing your seeds.

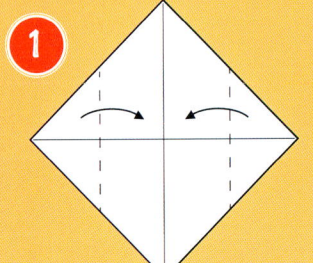
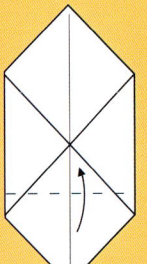
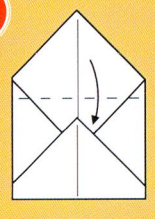

Poppies July 2024

Did you know?

You can dry out and save homegrown peas and beans, and seeds from peppers, courgettes, pumpkins and even apples. Try planting the seeds next year.

Bean plant seeds

Make Seed Bombs

Seed bombs are little balls of wildflower goodness. Just throw them onto a patch of soil and let nature take over!

You will need:
- A mixing bowl
- 3 cups of clay powder
- 4 cups of potting compost
- Wildflower seeds
- Water

1. Add the clay powder, compost and seeds to the bowl. Stir by hand to mix them up.

2. Add half a cup of water to the bowl. Start to knead the mixture. Carefully add more water, kneading until your mixture feels like dough.

3. Roll the mixture into balls, about the size of a table tennis ball.

4. Place your seed bombs in a sunny spot to dry.

5. Get ready to launch them. Throw or drop the seed bombs onto bare patches of soil in your garden or school playground – and leave them to grow!

Clay · Seeds · Seed bomb · Compost

Ask if you can start a "Growing Green" project at your school. Share this book with your teacher and friends. And never forget – EVERY ACTION COUNTS!

GLOSSARY

bulb
An underground plant part that stores food. A bulb stays underground all year. Many bulbs grow and flower in spring.

habitat
A place where plants, animals and other living things make their home. Forests, the ocean and gardens are all types of habitats.

compost
The brown, crumbly soil made by recycling food scraps and garden waste. Compost is full of nutrients that help plants grow.

herbs
Types of plants grown to add flavour in cooking, for use in medicine or for their scent. Lavender, thyme, basil and mint are all herbs.

microbe
A microscopic living thing. Some microbes are helpful and some, such as germs, can be harmful.

moss
Tiny plants that often grow on rocks and trees. They grow close together and create a carpet-like mass.

nectar
A sweet, sugary liquid that's produced by flowers. Insects such as bees and butterflies feed on nectar.

nutrients
Substances needed by living things to help them live and grow. Plants take in nutrients from the soil with their roots.

pollen
A fine, powdery dust that's produced by flowers. Pollen helps flowers produce seeds and is a food for beetles, bees and many other animals.

pollinate
To carry pollen from flower to flower. Insects, birds, bats and other animals help pollinate plants.

seed
A tiny part of a plant that contains all the material needed to grow a new plant.

seedling
A very young plant. Seedlings grow from seeds.

weed
A wild plant. Weeds are often tough and fast growing.

31

INDEX

B
bees 4, 9, 13, 15, 20, 23, 24, 26
beetles 4, 22
birds 7, 19, 24, 26
Bring in the Birds 19
bulbs 20–21
butterflies 4, 9, 21, 22, 25

C
Care for a Plant 5
Clever Clover 9
Compost and compost heaps 10–11, 16

E
environment 14, 25, 26

F
flowers 4, 9, 13, 15, 18, 20–21, 22–23, 24, 28–29

G
Grow a Mighty Oak Tree 27
Grow Your Own Strawberries 15
growing food 14–15, 16–17, 28–29

H
habitats 8, 19, 22, 24–25
herbs 14, 22

I
insects 4, 9, 10, 13, 14–15, 18–19, 20–21, 22–23, 24–25, 26

L
leftover food 10–11, 16–17
Let's Make Compost 11

M
Make a Bulb Lasagna 21
Make a Mini Moss Habitat 25
Make a Recycled Water Sprinkler 13
Make Seed Bombs 29

N
nectar 4, 9, 18, 20–21, 22–23, 24

P
pollen 4, 9, 18, 20, 22–23, 24, 26
pollinators 4, 15, 22, 24

R
Recycled Plant Labels 7
reuse and recycle 6–7

S
seeds 4, 6–7, 9, 14, 17, 22–23, 24, 27, 28–29
Slice-and-Grow Tomato 17
soil 5, 8–9, 10–11, 24
Super Giant Sunflowers 23

T
trees 26–27

W
water 4, 8–9, 12–13
weedkillers 5, 8
weeds 5, 11, 14, 24
wild areas 24–25
worms 9, 10, 24